STECK-VAUGHN

StartSmart®

Connecting Learning to Life

Math Strategies

Consultants

Jim Ford
Training Specialist
Center for Literacy Studies
University of Tennessee
Knoxville, Tennessee

Daniele D. Flannery, Ph.D.
Associate Professor of Adult Education
Coordinator: Adult Education D.Ed. Program
The Pennsylvania State University
Capital Campus—Harrisburg

Barbara Tondre-El Zorkani
Educational Consultant
Adult and Workforce Education
Austin, Texas

www.HarcourtAchieve.com
1.800.531.5015

Acknowledgments

Staff Credits

Executive Editor: Ellen Northcutt
Supervising Editor: Julie Higgins
Editor: Sharon Sargent
Director of Design: Scott Huber
Associate Director of Design: Joyce Spicer
Designer: Jim Cauthron
Production Manager: Mychael Ferris
Production Coordinator: Paula Schumann
Image Services Coordinator: Ted Krause
Senior Technical Advisor: Alan Klemp
Electronic Production Specialist: David Hanshaw

Cover Illustration

Joan Cunningham

ISBN 0-7398-6019-4

Printed in the United States of America

6 7 8 9 10 11 1421 15 14 13 12 11 10

4500239716

Contents

To the Learner

Congratulations! You have taken an important step as a lifelong learner. You have made the decision to take charge of your learning by improving your math skills. Steck-Vaughn's *Start Smart Math Strategies* will introduce you to a variety of strategies that will help you avoid common math errors and use math more confidently in your daily life.

Math is an active process. There is much more to math than just remembering formulas. To improve your math skills, you need to learn how to recognize and solve math problems. The strategies in this book will help you do just that. Some strategies will work better for you than others; that's fine. Try all the strategies and decide which ones work best for you.

As you work through this book be sure to:

- Fill out the **Identify Your Math Habits** chart on page 5. This will give you an idea of what your current math habits are.
- Write down your goals in the **Set Your Math Goals** web on page 7. Setting goals and checking your progress help you to stay motivated.
- As you read, check out the **Tips.** They provide a slightly different approach to the strategies.
- Complete the **Think About It** activities. These provide you with the opportunity to think about what you have learned. You'll also see a connection to everyday life.
- Create a **Journal.** In your journal you can write about your learning in your own words. This book is divided into five topics. At the end of each topic, you will have the opportunity to write in your journal. Writing thoughts in your own words will help you remember what you've learned and track your progress.
- Review what you have learned by completing **What Works for You?** on page 45.

Identify Your Math Habits

This chart can help you learn about your current math habits. As you use the strategies in this book, you may notice that your habits are changing.

For each statement, circle A (always), S (sometimes), or N (never).

When I do math, I . . .	Always	Sometimes	Never
use place value to understand numbers.	A	S	N
understand number lines.	A	S	N
compare whole numbers to understand value.	A	S	N
round numbers to estimate.	A	S	N
identify clues in word problems.	A	S	N
do math mentally first, if possible.	A	S	N
check my answers.	A	S	N
line up digits by place value and regroup when necessary.	A	S	N
apply the properties of addition.	A	S	N
apply the properties of subtraction.	A	S	N
apply the properties of multiplication.	A	S	N
apply the properties of division.	A	S	N
account for remainders in division.	A	S	N

Set Your Math Goals

For the things of this world cannot be made known without a knowledge of mathematics.

Roger Bacon, English philosopher,
scientist, and mathematician
(1214–1294)

When you buy a car, follow a recipe, or decorate your home, you're using math. People around the world have used the strategies covered in this book for over thousands of years. Using math strategies can help you make important decisions and perform everyday tasks. Math can help us shop wisely, buy the right insurance, or even follow a baseball game.

Why is learning math skills important to you? Explain in your own words.

....

....

....

....

....

....

Consider how improving your math skills could help you in these areas of your life.

- **Family:** How could improved math skills help make your family life better?
- **Work:** How could it help you in your job?
- **Community:** How could it help you serve your community?
- **Self:** How could it help you continue your education or explore a new interest?

Write your math goals in the spaces below. Think about how becoming better in math will affect your family, your work, your community, and yourself.

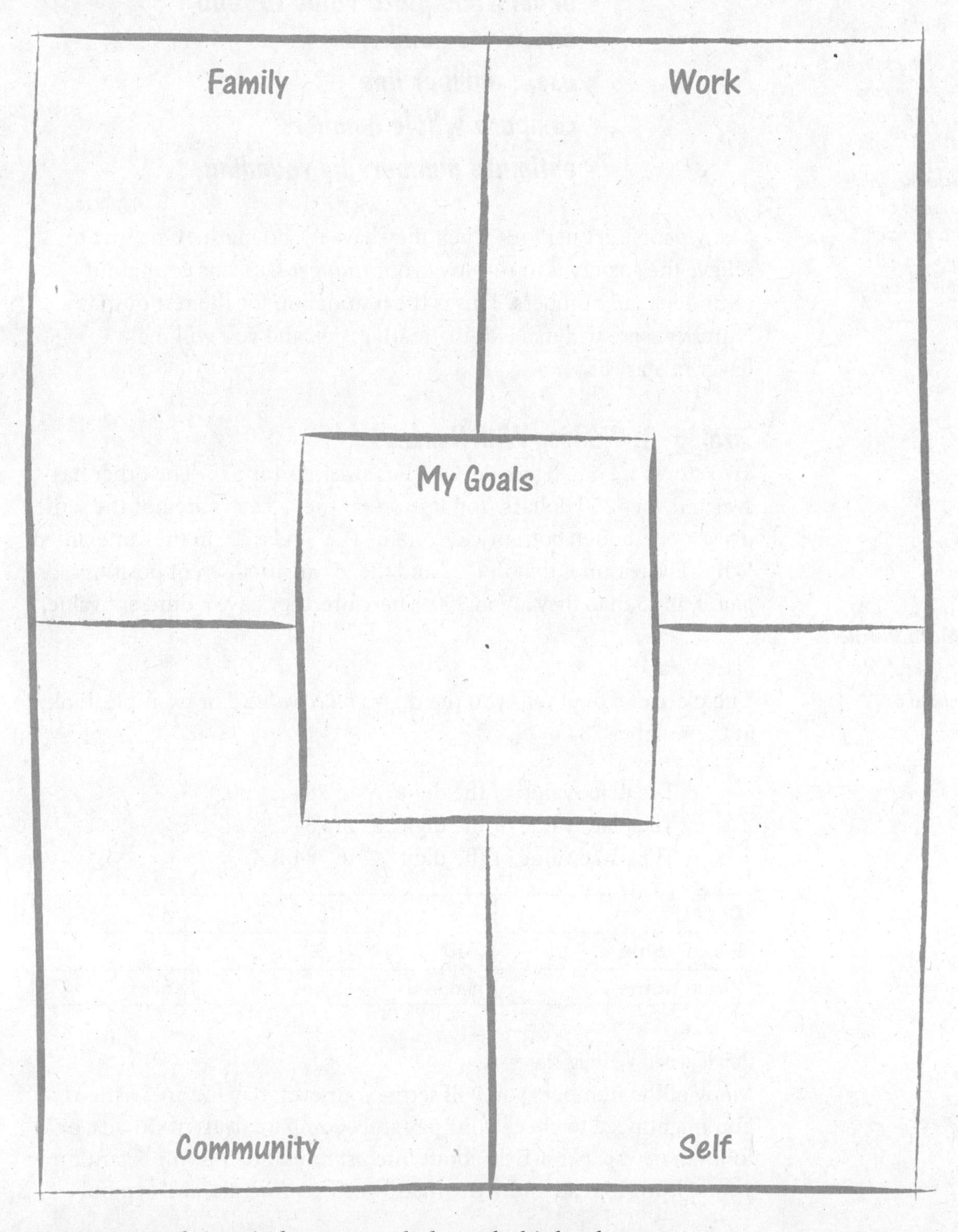

Keep your goals in mind as you work through this book.

numeral a symbol that represents a number. The numerals in our number system are 0, 1, 2, 3, 4, 5, 6, 7, 8, and 9.

digit each of the numerals that make up a number. The number 137 has three digits: 1, 3, and 7.

place value the value of a digit, based on its position in a number. The number 7 in 137 is in the ones place.

Topic 1: How to Make Sense of Numbers

In this section, you will learn how to:

- ***understand place value to read and write numbers***
- ***use a number line***
- ***compare whole numbers***
- ***estimate numbers by rounding***

Many people get nervous when they have to "do math." One way to relieve the anxiety is to improve your *number sense,* or your ability to understand numbers. This is the foundation for the rest of math. Number sense will make doing math easier, and you will make fewer mistakes.

Strategy 1: Working With Numbers

Two stores are having a sale. One has sweaters for $25. The other has sweaters for $250 dollars. You know that the sweaters are not the same price, even though both prices contain a "2" and a "5" in the same order. Why? The reason is that the "2" and the "5" are in different positions, or *places,* in 25 than they are in 250. Therefore, they have a different value.

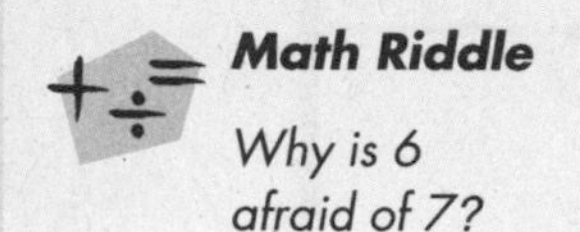

Math Riddle

Why is 6 afraid of 7?

Because 7 ate 9. (7, 8, 9) Share this riddle with your kids or friends.

Place Value

The place of a digit tells you the digit's *place value.* For example, look at the number 784.

The place value of the digit "7" in 784 is 700.
The place value of the digit "8" in 784 is 80.
The place value of the digit "4" in 784 is 4.

Digits	7	8	4
Place value	700	80	4
Place name	hundreds	tens	ones

Reading and Writing Numerals

Many of the numbers you will see in your everyday life are written. You might need to check your pay stub, compare deals in car ads, or follow a newspaper article about interest rates. So, it is important for you to learn to read and write math words, numerals, and symbols.

You can use a place value chart to read and write numbers written in numerals. For example, write the number 784 in words.

Words	seven hundred	eighty	four
Digits	7	8	4
Place value	700	80	4
Place name	hundreds	tens	ones

The number 784 is written in words as "seven hundred eighty-four." Do not use the word *and* to combine the words for whole numbers.

A place value chart can also help you read and write numbers written in words. Write the number "three thousand five hundred sixty-seven" in numerals.

Digits	3	5	6	7
Place value	3,000	500	60	7
Place name	thousands	hundreds	tens	ones
Words	three thousand	five hundred	sixty	seven

The number "three thousand five hundred sixty-seven" is written in numerals as 3,567.

Try It Yourself

Write the place name and the place value for the 5 in each of these numbers.

1. 825 ones ______ 5 ______

2. 9,350 ______ ______

3. 20,506 ______ ______

4. 5,890 ______ ______

TIP

Four-digit numbers can be written with or without a comma: 4999 or 4,999.

Write each number in words.

5. 21 twenty-one

7. 65 ______

6. 13,000 ______

8. 1,340 ______

Write each number in numerals.

9. twenty-five 25

10. ninety-nine ______

11. three hundred seven ______

12. one thousand, one hundred twenty ______

Check your answers on page 46.

Math Riddle

What happened in 1961 that will not happen again for over 4,000 years?

The year's date reads the same when turned upside down. This will not happen again until 6009.

Reading and Writing Symbols

Numbers and math symbols are everywhere. They are found on road signs, pay checks, and advertisements. Here are some common math symbols that you should learn to read and write.

$+$	add, plus	$=$	equals, is
$-$	subtract, minus	$\neq$	is not equal to, does not equal
$\times$	multiply, times	$>$	is greater than
$\div$	divided by	$<$	is less than

Example The examples below show how symbols can be combined with numerals to make math sentences.

$14 + 12 = 26$	Fourteen **plus** twelve **equals** twenty-six.
$30 > 29$	Thirty **is greater than** twenty-nine.
$29 < 30$	Twenty-nine **is less than** thirty.
$5 \times 7 = 35$	Five **times** seven **is** thirty-five.
$9 - 9 = 0$	Nine **minus** nine **equals** zero.
$10 \div 2 = 5$	Ten **divided by** two **equals** five.
$8 \neq 15$	Eight **is not equal to** fifteen.

Try It Yourself

Write the correct word, phrase, or symbol. Check your answers on page 46.

1. $\neq$ not equal to ______ **4.** $=$ ______

2. $\times$ ______ **5.** $<$ ______

3. divide ______ **6.** is greater than ______

If the math sentence is written in words, write it in symbols and numerals. If the sentence is written in symbols and numerals, write it in words.

7. $9 + 1 = 10$ Nine plus one equals ten.

8. Three times five equals fifteen. ______

9. $6 - 2 = 4$ ______

10. $45 \neq 46$ ______

TIP

Circle the symbol before you begin to work. Calling attention to the symbol will help you focus on what you need to do.

Think About It

How did you use numbers and math symbols this past week?

....

....

number line *a line with equally spaced points that are labeled with equally spaced numbers.*

negative number *a number that is less than zero.*

positive number *a number that is greater than zero.*

Strategy 2: Using a Number Line

Numbers less than zero are negative numbers. Numbers greater than zero are positive. Zero is neither negative nor positive. On a number line, positive numbers are to the right of zero and negative numbers are to the left.

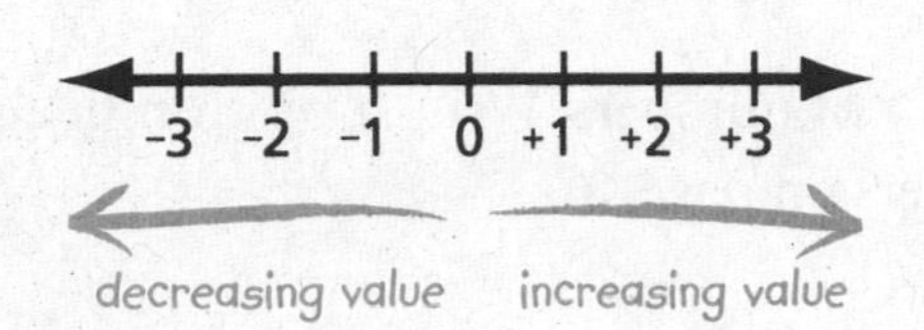

Sometimes drawing a number line can help you solve math problems. A number line shows the relationship between numbers. Numbers increase in value to the right and decrease in value to the left. The arrows at either end of the line show that numbers continue in both directions, less than −3 and greater than 3.

TIP

A positive number can be written with or without a positive sign (+). Negative numbers must always have a negative sign (−) in front of the number.

Try It Yourself

Where would you find the following numbers on a horizontal number line? Write **left of zero** or **right of zero.** Check your answers on page 46.

1. 12 right of zero **2.** −14 ____________ **3.** 45 ____________

Think About It

What kinds of math problems do you solve where drawing a number line or picture might help?

....

....

Strategy 3: Comparing Whole Numbers

Would you rather pay $589 or $489 for the same TV? To answer the question, you would compare the two numbers.

whole numbers *the positive numbers and zero: 0, 1, 2, 3 . . .*

Here are three rules for comparing whole numbers.

TIP
Think of the "greater than" and "less than" symbols as arrows that always point to the smaller number.

Rule 1: A whole number with fewer digits is less than a whole number with more digits. For example, 99 is less than 367 because 367 has three digits and 99 has only two digits. **99 < 367**

Rule 2: If two whole numbers have the same number of digits, compare the digits with the greatest place value. For example, 589 is greater than 489 because 5 is greater than 4. **589 > 489**

Rule 3: If the digits with the greatest place value are equal, move to the right and compare the digits. For example, 502 is less than 520 because 0 is less than 2. **502 < 520**

Try It Yourself

Compare each pair of numbers. Write > or < on the line provided. Check your answers on page 46.

1. 3,057 __<__ 3,129 **3.** 57 ____ 29 **5.** 47 ____ 370

2. 1,241 ____ 1,291 **4.** 983 ____ 984 **6.** 16,500 ____ 16,050

Think About It

How do you use comparisons to make choices?

....

....

Strategy 4: Estimating Numbers by Rounding

Rounding gives you numbers that are easier to work with. You can round numbers when you do not need to find an exact answer. For example, Julia brought $30 to the store. Does she have enough money to buy a shirt for $19 and a pair of pants for $18?

rounding *expressing a number as an approximate value; often done to make calculations easier.*

Example Round 6,382 and 6,782, and 9,782 to the nearest thousand.

Step 1: Underline the digit in the thousands place. Then look at the digit to the right.	6,382 ↑	6,782 ↑	9,782 ↑
Step 2: Use one of these rules:			
Rule 1: If the digit is less than 5, **round down** by changing all digits to the right of the underlined digit to 0.	$3 < 5$ 6, 382 rounds down to 6,000		
Rule 2: If the digit is 5 or greater, **round up** by adding one to the underlined digit and then changing all digits to the right to 0.		$7 > 5$ 6, 782 rounds up to 7,000	
Rule 3: If the digit is 5 or greater and the underlined digit is 9, **round up** by adding one to the digit to the left of the underlined digit and then changing all digits to the right to 0.			$7 > 5$ and underlined digit is 9, 9,782 rounds up to 10,000

Try It Yourself

For each number, decide whether to **round up** or **round down.** Name the underlined place. Then round to the underlined place.

1. 4,567 hundreds; 4,600

2. 665 ______

3. 302 ______

4. 12,599 ______

Think About It

When working with numbers, when could you estimate by rounding and when do you need an exact answer?

....

....

When You Make Sense of Numbers, you

- Use place value to understand numbers.
- Use a number line.
- Compare whole numbers to understand value.
- Estimate numbers by rounding.

Math Riddle

What comes next in this letter sequence: Z, O, T, T, F, F, S, S, E?

*The next letter is N: **z**ero, **o**ne, **t**wo, **t**hree, **f**our, **f**ive, **s**ix, **s**even, **e**ight, **n**ine.*

Journal

In your journal, describe how you will use the strategies taught in Topic 1.

Next, look at the goals you set on page 7. Note your progress for each of your goals. If you've achieved a goal, put a check next to it and congratulate yourself!

addition combining two or more amounts to get a single total

sum the answer to an addition problem

Topic 2: Addition

In this section, you will learn how to:

- ***recognize and apply properties of addition***
- ***add numbers and regroup amounts as necessary***
- ***solve word problems with addition***
- ***add numbers using mental math***
- ***estimate to check sums***

Everyday tasks that may require addition include figuring out a budget, estimating the cost of groceries, and balancing your checkbook.

Strategy 1: Recognize and Apply Addition Properties

Here are three facts about addition that you need to know:

1. The Identity Property

Adding zero to a number does not change the value of the number.

Example $4 + 0 = 4$

$4 = 4$

2. The Commutative Property

When you add numbers, changing the order of the numbers does not change the sum.

Example $2 + 3 = 3 + 2$

$5 = 5$

3. The Associative Property

When you add a group of three or more numbers, changing the pair of numbers you add first does not change the sum.

Example $(1 + 2) + 3 = 1 + (2 + 3)$

$3 + 3 = 1 + 5$

$6 = 6$

Try It Yourself

Apply the identity, commutative, or associative property of addition to make each number sentence true. State which property you use. Check your answers on page 46.

1. 7 + 8 = 8 + __7__ Property: commutative
2. 0 + 91 = ____ Property: ____________
3. 5 + (7 + 9) = (5 + ____) + 9 Property: ____________
4. 4,288 + ____ = 4,288 Property: ____________
5. 20 + 30 + 40 = ____ + 30 + 20 Property: ____________
6. (44 + 55) + ____ = 44 + (55 + 67) Property: ____________

Work with a Partner

Try this adding game. Start with the target number 43. The first player chooses a whole number from 1 to 7. You and your partner take turns to add a whole number from 1 to 7 to the running total. The one who hits the target of 43 wins the game. To change the game, choose a new target or a new range of numbers to add on.

Think About It

What names could you give to the three properties of addition to help you remember them?

....

....

Strategy 2: Use Place Value to Add

When you add numbers, you may **only combine digits with like place values.** You must add ones to ones, tens to tens, hundreds to hundreds, etc.

For help understanding place value, review page 8.

Basic Addition

To add two or more numbers, first line up the numbers in a column by place value. Then add the digits in each place.

Example To add 23 and 15, follow the steps below.

Step 1: Align by place value.

	tens	ones
	2	3
+	1	5

Step 2: Add the ones.

	tens	ones
	2	3
+	1	5
		8

Step 3: Add the tens.

	tens	ones
	2	3
+	1	5
	3	8

23 + 15 = 38.

Example Add 502 and 94.

Step 1: Align by place value.	Step 2: Add the ones.	Step 3: Add the tens.	Step 4: Add the hundreds.
hundreds tens ones	hundreds tens ones	hundreds tens ones	hundreds tens ones
5 0 2	5 0 2	5 0 2	5 0 2
+ 9 4	+ 9 4	+ 9 4	+ 9 4
	6	9 6	5 9 6

502 + 94 = 596.

Try It Yourself

Write each addition problem in a column and solve. Check your answers on page 46.

1. 16 + 10 **2.** 23 + 4 **3.** 189 + 710 **4.** 842 + 44 **5.** 35 + 12 + 11

```
  16
+ 10
  26
```

Work with a Group

With your group, solve this math riddle. *Write a three-digit number using the digits 3, 6, and 9. The hundreds digit must be the sum of the tens and one digits. The ones digit must be greater than the tens digit.* Check your answer on page 46.

Addition with Regrouping

regroup *move amounts between place values*

What happens when the sum of the digits in a place value is greater than 9? To find the answer, try adding 64 and 28.

Example

```
 tens ones
   6   4
 + 2   8
      12?  ←  Add the ones. 12 ones = 10 ones + 2 ones
                                   = 1 ten + 2 ones
```

When you add the ones you get 12 ones. You know that 12 ones is the same as 1 ten and 2 ones. You can write the 2 ones in the ones place. However, tens do not belong in the ones place. You must move, or **regroup,** the 1 ten to the tens place. Then you can add the tens.

	tens	ones	
	1		← Regroup the 1 ten to the tens place.
	6	4	
+	2	8	
		2	← Write the 2 ones in the ones place.

	tens	ones	
	1		
	6	4	
+	2	8	
	9	2	← Add the tens.

64 + 28 = 92.

Sometimes, you will need to regroup more than once.

Example Add 75 and 88.

	hundreds	tens	ones	
		7	5	
+		8	8	
			13?	← Add the ones. 13 ones = 10 ones + 3 ones = 1 ten + 3 ones

	hundreds	tens	ones	
		1		← Regroup the 1 ten to the tens place.
		7	5	
+		8	8	
			3	← Write the 3 ones in the ones place.

	hundreds	tens	ones	
		1		
		7	5	
+		8	8	
		16?	3	← Add the tens. 16 tens = 10 tens + 6 tens = 1 hundred + 6 tens

	hundreds	tens	ones	
	1	1		← Regroup the 1 hundred to the hundreds place.
		7	5	
+		8	8	
		6	3	← Write the 6 tens in the tens place.

	hundreds	tens	ones	
	1	1		
		7	5	
+		8	8	
	1	6	3	← Add the hundreds.

75 + 88 = 163.

TIP

10 ones = 1 ten
10 tens = 1 hundred
10 hundreds = 1 thousand

TIP

Cover up all the columns except the one you are working on.

Try It Yourself

Write each addition problem in a column and solve. Check your answers on page 46.

1. 11 + 9 **2.** 392 + 84 **3.** 64 + 59 **4.** 480 + 566

```
  1
  11
+  9
  20
```

Start Smart Challenge

How fast can you add 9,191 and 1,919?

Strategy 3: Look for Clues in Addition Word Problems

When you read a word problem, terms like *altogether, combined, in all,* and *total* are clues telling you that you may need to use addition to solve the problem.

Example Jack buys 2 pounds of nails and 5 pounds of screws. What is the total number of pounds of hardware Jack buys?

The problem asks you to find a total. The word *total* is a clue that this is an addition problem. To solve it, add the weight of the nails (2 pounds) and the weight of the screws (5 pounds).

TIP
Talk to yourself as you solve word problems. This will help you think through the problem.

2 + 5 = 7 Jack buys a total of 7 pounds of hardware.

Try It Yourself

Circle the word or words in each problem that suggest solving with addition. Write and solve an addition number sentence to answer each problem. Check your answers on page 46.

1. Peggy walks for 20 minutes in the morning and for 15 minutes in the afternoon. How long does she walk (altogether)?

```
  20
+ 15
  35
```

Peggy walks for 35 minutes altogether.

2. Ramona typed 8 pages yesterday and 13 pages today. How many pages did she type in all?

....

3. Curtis lifts a 25-pound weight in each hand at the same time. What is the combined weight Curtis lifts?

....

Work With a Partner

Create a short word problem that a student could solve with addition. The solution should be $25. Share your word problem with the class.

Think About It

What other words might tell you that you need to use addition to solve a word problem?

....

....

Math Riddle

Find four odd numbers that follow each other and whose sum is 4,456.

1,111; 1,113; 1,115; 1,117

Strategy 4: Count to Ten to Do Mental Math

Suppose you need to add the prices of four items at the store. The prices are $7, $4, $3, and $6. You probably cannot sit down in the middle of the store with a pencil and paper to do the calculation. You will need to add the prices "in your head." This is called **mental math.**

Make it easier on yourself by looking for numbers where the **ones** place values add up to 10.

Example

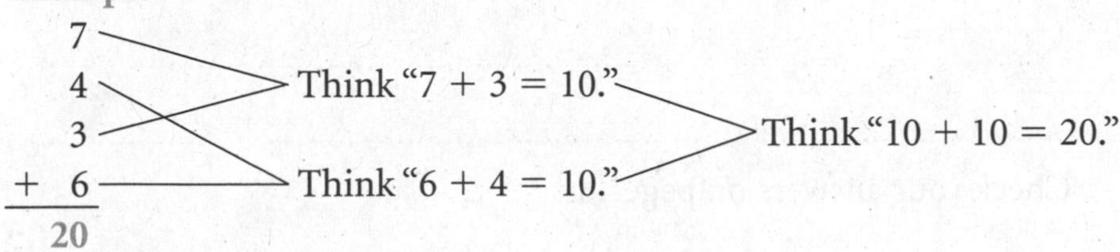

Try It Yourself

Add each column using mental math. Check your answers on page 46.

1.	**2.**	**3.**	**4.**	**5.**
5	2	22	10	35
2	8	31	39	79
5	+ 50	28	+ 31	25
+ 8		+ 9		+ 81
20				

Think About It

When do you use mental math?

....

....

For help understanding rounding, review page 13.

Strategy 5: Estimate to Check Sums

After you add, always check that the sum is reasonable. One way to check a sum is to round the numbers you added and then estimate the total. If your calculated sum and the estimate are close, your sum is probably reasonable.

Example Bonnie wants to buy a video for $28 and a CD for $9. Bonnie decides that the two items will cost $37. Is Bonnie's sum reasonable?

Estimated Sum		**Compare to Calculated Sum**
$28 rounds to $30	$30	$37 is close to $40
$9 rounds to $10	+ 10	$37 is a reasonable sum.
	$40	

Try It Yourself

1. Carol must keep track of the miles she drives her work truck. She drives 39 miles on Monday and 23 miles on Tuesday. Carol tells her supervisor that she drove a total of 62 miles during the two days. Is Carol's sum reasonable? Explain.

....

....

2. Jim spends $12 for a shirt and $19 for pants to wear to work. He calculates that the total cost is $51. Is Jim's answer reasonable? Explain.

....

....

Check your answers on page 46.

Think About It

What is the most significant thing you have learned about math today?

....

When You Add Numbers

- Apply the properties of addition.
- Line up digits by place value and regroup when necessary.
- Identify clues that tell you to add in a word problem.
- Look for digits that add up to ten to do mental math.
- Estimate to check your answer.

Journal

In your journal, describe what images come to mind when you think about jobs involving math.

Next, look at the goals you set on page 7. Note your progress for each of your goals. If you have achieved a goal, put a check next to it and congratulate yourself!

Topic 3: Subtraction

In this section, you will learn how to:

- ***recognize and apply properties of subtraction***
- ***subtract numbers and regroup amounts as necessary***
- ***solve word problems with subtraction***
- ***subtract numbers using mental math***
- ***check differences using estimation and addition***

Everyday tasks that may require subtraction include comparing prices, making change, and tracking weight loss.

subtraction *taking away a quantity from another quantity*

difference *the answer to a subtraction problem*

Strategy 1: Recognize and Apply Subtraction Properties

Here are two facts about subtraction that you need to know:

1. The Identity Property

Subtracting zero from a number does not change the value of the number.

Example $3 - 0 = 3$

$3 = 3$

2. The Zero Property

Subtracting a number from itself always equals zero.

Example $4 - 4 = 0$

$0 = 0$

Try It Yourself

Apply the identity or zero property of subtraction to make each number sentence true. State which property you use. Check your answers on page 46.

1. $6 - 0 =$ 6 Property: identity

2. $2 -$ ____ $= 0$ Property: ____________

3. ____ $- 8 = 0$ Property: ____________

4. $7 -$ ____ $= 7$ Property: ____________

Work With a Partner

On your own paper, draw an example for both of the subtraction properties without naming which property you are showing. Exchange examples with a partner and identify which one matches each property.

Think About It

Why does subtracting zero from a number not change the value of the number?

....

....

For help understanding place value, review page 8.

Strategy 2: Use Place Value to Subtract

To subtract, first identify the greater number. Write the numbers in a column with the lesser number **under** the greater number.

Basic Subtraction

When you subtract, you may **only subtract digits with like place values.** You must subtract ones from ones, tens from tens, hundreds from hundreds, etc. Just like addition, you must line up the numbers in a column by place value.

Example To subtract 4 from 19, follow the steps below.

Step 1: Write the lesser number under the greater number and line up by place value.

	tens	ones
	1	9
−		4

Step 2: Subtract the ones.

	tens	ones
	1	9
−		4
		5

Step 3: Subtract the tens.

	tens	ones
	1	9
−		4
	1	5

19− 4 = 15.

Example Subtract 23 from 648.

Step 1: Write the lesser number under the greater number and line up by place value.

	hundreds	tens	ones
	6	4	8
−		2	3

Step 2: Subtract the ones.

	hundreds	tens	ones
	6	4	8
+		2	3
			5

Step 3: Subtract the tens.

	hundreds	tens	ones
	6	4	8
+		2	3
		2	5

Step 4: Subtract the hundreds.

	hundreds	tens	ones
	6	4	8
+		2	3
	6	2	5

648 − 23 = 625.

Try It Yourself

Write each subtraction problem in a column and solve. Check your answers on page 46.

1. 58 − 6 **2.** 21 − 10 **3.** 79 − 55 **4.** 841 − 41 **5.** 976 − 324

$$\begin{array}{r} 58 \\ -\ \ 6 \\ \hline 52 \end{array}$$

Subtraction with Regrouping

What happens if the digit being subtracted is greater than the digit above it? To answer this question, try subtracting 8 from 42.

For help understanding regrouping, review page 16.

Example

	tens	ones
	4	2
−		8
		?

← You cannot subtract 8 ones from 2 ones.

You know that 1 ten is the same as 10 ones. Take 1 ten away from the tens place and *regroup* it to the ones place as 10 ones.

Step 1: Regroup 1 ten to the ones place (2 ones + 10 ones = 12 ones).

	tens	ones
	3	12
	~~4~~	~~2~~
−		8

Step 2: Subtract the ones.

	tens	ones
	3	12
	~~4~~	~~2~~
−		8
		4

Step 3: Subtract the tens.

	tens	ones
	3	12
	~~4~~	~~2~~
−		8
	3	4

42 − 8 = 34.

TIP
Sometimes, you will need to regroup more than once.

Example Subtract 57 from 846.

	hundreds	tens	ones
	8	4	6
−		5	7
			?

← You cannot subtract 7 ones from 6 ones.

Step 1: Regroup 1 ten to the ones place (6 ones + 10 ones = 16 ones).

	hundreds	tens	ones
		3	16
	8	~~4~~	~~6~~
−		5	7

Step 2: Subtract the ones.

	hundreds	tens	ones
		3	16
	8	~~4~~	~~6~~
−		5	7
			9

	hundreds	tens	ones
		3	16
	8	~~4~~	~~6~~
−		5	7
		?	9

← You cannot subtract 5 tens from 3 tens.

Step 3: Regroup 1 hundred to the tens place (3 tens+ 10 tens = 13 tens).

	hundreds	tens	ones
		13	
	7	~~3~~	16
	~~8~~	~~4~~	~~6~~
−		5	7
			9

Step 4: Subtract the tens.

	hundreds	tens	ones
		13	
	7	~~3~~	16
	~~8~~	~~4~~	~~6~~
−		5	7
		8	9

Step 5: Subtract the hundreds.

	hundreds	tens	ones
		13	
	7	~~3~~	16
	~~8~~	~~4~~	~~6~~
−		5	7
	7	8	9

846 − 57 = 789.

Here's how to regroup when the digit you want to "borrow" from is 0.

Example Subtract 64 from 301.

	hundreds	tens	ones
	3	0	1
−		6	4
			?

← You cannot subtract 4 ones from 1 ones. However, there are no tens to regroup to the ones place.

Step 1: Regroup 1 hundred as 10 tens to the tens place. (0 tens + 10 tens = 10 tens).

	hundreds	tens	ones
	2	10	
	~~3~~	~~0~~	1
−		6	4

Step 2: Regroup 1 ten to the ones place. (1 ones + 10 ones = 11 ones).

	hundreds	tens	ones
		9	
	2	~~10~~	11
	~~3~~	~~0~~	~~1~~
−		6	4

Step 3: Subtract the ones, then the tens, and then the hundreds.

	hundreds	tens	ones
		9	
	2	~~10~~	11
	~~3~~	~~0~~	~~1~~
−		6	4
	2	3	7

301 − 64 = 237.

Try It Yourself

Write each subtraction problem in a column and solve. Check your answers on pages 46–47.

1. 71 − 28 **3.** 115 − 92 **5.** 651 − 277 **7.** 805 − 7

$$\begin{array}{r} \scriptstyle 6\ 11 \\ \not{7}\not{1} \\ -\ 2\ 8 \\ \hline 4\ 3 \end{array}$$

2. 436 − 81 **4.** 332 − 199 **6.** 400 − 35 **8.** 503 − 73

Start Smart Challenge

Subtract 98,765,432 from 99,999,999.

TIP

Use graph paper to keep numerals lined up. Allow one numeral per square.

Strategy 3: Look for Clues in Subtraction Word Problems

When you read a word problem, words and phrases like *difference, how much less (smaller, slower, etc.), how much more (greater, taller, etc.), left over,* and *remain* are clues to tell you that you may need to use subtraction to solve the problem.

Example Penny's daughter was 58 inches tall last year on her birthday. Now, her daughter is 63 inches tall. How much taller is Penny's daughter now than on her last birthday?

The problem asks you to find the change in the daughter's height. The phrase *how much taller* is a clue that this is a subtraction problem. To solve it, subtract the daughter's height last year (58 inches) from her height now (63 inches).

$$\begin{array}{r} \scriptstyle 5\ 13 \\ \not{6}\ \not{3} \\ -\ 5\ 8 \\ \hline 5 \end{array}$$

Penny's daughter is 5 inches taller.

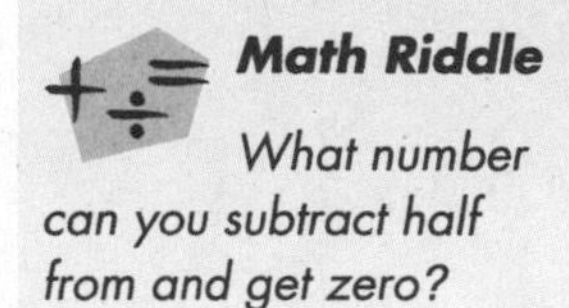

Math Riddle

What number can you subtract half from and get zero?

The number 8 because it is made of 2 zeros one on top of the other.

Try It Yourself

Circle the word or words in each problem that suggest solving with subtraction. Write and solve a subtraction number sentence to answer each problem. Check your answers on page 47.

1. Last week David spent $40 on groceries. This week he spent $52 on groceries. What is the (difference) between David's spending last week and this week? **52 − 40 = 12; David spends $12 more this week, or David spent $12 less last week.**

2. Melanie has a 24-ounce can of tomatoes. She uses 16 ounces for a recipe. How many ounces of tomatoes does Melanie have left over?

....

....

3. Alfredo weighs 175 pounds. After going on a diet- and exercise-plan, he now weighs 150 pounds. How much less does Alfredo weigh now?

....

....

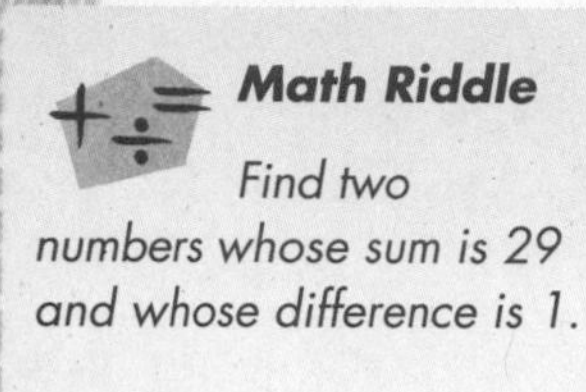

Math Riddle

Find two numbers whose sum is 29 and whose difference is 1.

14, 15

4. A video is 105 minutes long. If the video has played for 59 minutes, how many minutes remain? ....

....

Work With a Partner

Create a short word problem that a student could solve with subtraction. The solution should be $36. Share your word problem with the class.

Think About It

What other words or phrases might tell you that you need to use subtraction to solve a word problem?

....

....

Strategy 4: Use Ten to Do Mental Math

Subtracting numbers like 10, 20, 30, 40, and so on, is usually pretty easy. You can do it "in your head" because you do not have to regroup to subtract 0.

Example Subtract 24 from 82.

Step 1: $\begin{array}{r} 82 \\ -\ 24 \\ \hline ? \end{array}$ ← Think "The next ten **after** 24 is 30."

Step 2: $\begin{array}{r} 82 \\ -\ 30 \\ \hline 52 \end{array}$ ← Subtract 30.

Step 3: $\begin{array}{r} 52 \\ +\ 6 \\ \hline 58 \end{array}$ ← You subtracted more than you needed to, so you need to add it back to get the correct answer. Think "24 $\underline{+\ 6}$ = 30, so I must add 52 $\underline{+\ 6}$ to the final answer."

82 − 24 = 58

Try It Yourself

Subtract using mental math. Check your answers on page 47.

1. 15	**2.** 65	**3.** 33	**4.** 22	**5.** 93
− 8	− 29	− 14	− 3	− 57
7				

Think About It

How do you know when to regroup in subtraction?

....

....

Strategy 5: Use Estimation or Addition to Check Differences

After you subtract, always check your answer. One way to check your answer is to use estimation.

Round the numbers you subtracted and then estimate the difference. If your calculated difference and the estimated difference are close, your calculated difference is probably reasonable.

For help understanding rounding, review page 13.

Example Cynthia has $61. She estimates that if she buys a drill for $48, she will have $13 left. Is Cynthia's difference reasonable?

Estimated Difference		Compare to Calculated Difference
$61 rounds to $60	$60	$13 is close to $10
$48 rounds to $50	− 50	$13 is a reasonable difference.
	$10	

TIP

When estimating the cost of several items, it is better to round up. That way, the bill won't be more than you expect. Remember that tax will be added as well.

Another way to check a difference is to use addition. Addition and subtraction are *inverse operations.* This means that you can use addition to "undo" subtraction and use subtraction to "undo" addition. Here's how to use addition to check the answer to a subtraction problem.

Example Richard calculates 42 − 19 = 23. Is this difference correct?

Richard started at 42 and got got to 23 by **subtracting 19.**	Can he start at 23 and get back to 42 by **adding 19?**
$\begin{array}{r} {}^{3}\not{4}\,{}^{12}\not{2} \\ -\ 1\ 9 \\ \hline 2\ 3 \end{array}$	$\begin{array}{r} {}^{1}\ \ \\ 2\ 3 \\ +\ 1\ 9 \\ \hline 4\ 2 \end{array}$ ← Yes. The difference is correct.

Try It Yourself

Use estimation to answer each question.

1. Last year Marco weighed 215 pounds. Now he weighs 187 pounds. Marco calculates that he has lost 28 pounds. Is this difference reasonable? Explain. ....

....

2. In one month, Kevin's family uses 476 kilowatts of electricity and Trent's family uses 603 kilowatts of electricity. Kevin says that his family used 157 fewer kilowatts than Trent's family. Is Kevin's difference reasonable? Explain. ....

....

Math Riddle

How do five and nine make two?

What comes five hours after nine o'clock? Two o'clock.

Use addition to check each difference. If the answer is incorrect, state the correct difference.

3.	**4.**	**5.**	**6.**	**7.**
43	94	75	400	350
− 27	− 85	− 38	− 261	− 72
16?	9?	33?	139?	288?

$$\begin{array}{r} ^{1} \\ 16 \\ +\,27 \\ \hline 43 \end{array}$$

Check your answers on page 47.

Think About It

How could you use subtraction to check a sum?

....

....

When You Subtract Numbers

- Apply the properties of subtraction.
- Align digits by place value and regroup when necessary.
- Identify clues that tell you to subtract in word problems.
- Move up to the next "ten" to do mental math.
- Check differences with estimation or addition.

Journal

In your journal, describe how you might teach someone to subtract 47 from 81.

Next, look at the goals you set on page 7. Note your progress for each of your goals. If you have achieved a goal, put a check next to it and congratulate yourself!

Topic 4: Multiplication

In this section, you will learn how to:

- ***model multiplication problems***
- ***recognize and apply properties of multiplication***
- ***multiply numbers and regroup as necessary***
- ***solve word problems with multiplication***
- ***multiply by 10, 100, and 1,000 using mental math***
- ***estimate to check products***

Multiplication is something we use every day. Calculating your salary, estimating distance driven, and keeping track of loan payments are all tasks that require multiplication.

multiplication *adding the same number multiple times*

product *the answer to a multiplication problem*

Strategy 1: Understanding Multiplication Problems

Multiplication is repeated addition. Or, in other words, multiplication is a fast way of adding a series of numbers. "5 × 3" is another way to say "start at 0 and add 5 three times," or "5 + 5 + 5."

You can use a picture or a number line to show what is going on in a multiplication problem and to find the product.

TIP
Make a stack of empty multiplication tables. Keep them handy so you can practice often.

Example Use a picture to model 5 × 3 and find the product.

$5 \times 3 = 5 + 5 + 5$

$5 \times 3 =$ [■■■■■] + [■■■■■] + [■■■■■]

$5 \times 3 = 15$

Example Use a number line to model 5 × 3 and find the product.

$5 \times 3 = 5 + 5 + 5$

$5 \times 3 =$ 0 1 2 3 4 5 6 7 8 9 10 11 12 13 14 15 16 17

$5 \times 3 = 15$

Try It Yourself

1. Create a picture to model 2 × 4 and find the product.

2. Create a number line to model 3 × 6 and find the product.

Check your answers on page 47.

Think About It

Which kind of model helps you most to understand multiplication? Why?

....

....

Strategy 2: Recognize and Apply Multiplication Properties

Here are three facts about multiplication that you need to know.

1. The Zero Property

Multiplying a number by 0 gives a product equal to 0

Example $0 \times 3 = 0$

$0 + 0 + 0 = 0$ ▢ + ▢ + ▢ = 0

$0 = 0$

2. The Identity Property

Multiplying a number by 1 does not change the value of the number.

Example $1 \times 4 = 4$

$1 + 1 + 1 + 1 = 4$ ▣ + ▣ + ▣ + ▣ = 4

or 0 1 2 3 4 5 6 7 8

$4 = 4$

3. The Commutative Property

When you multiply numbers, changing the order does not change the product.

Example $5 \times 3 = 3 \times 5$

$5 + 5 + 5 = 3 + 3 + 3 + 3 + 3$

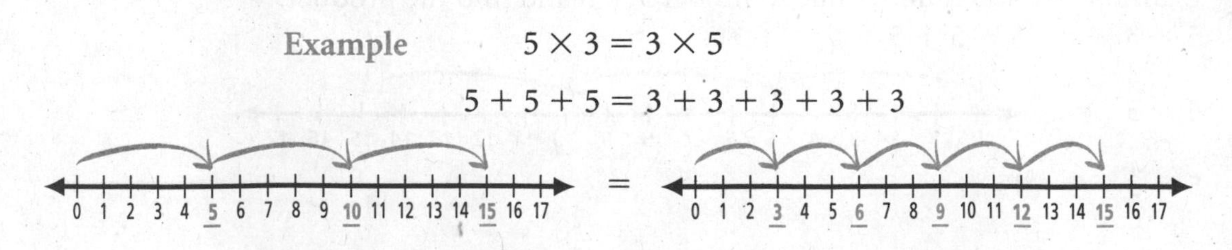

$15 = 15$

Try It Yourself

Apply the identity, zero, or commutative property of multiplication to make each number sentence true. State which property you use. Check your answers on page 47.

1. $3 \times 2 = 2 \times$ <u>3</u> Property: <u>commutative</u>
2. $14 \times$ ____ $= 0$ Property: __________
3. $85 \times$ ____ $= 85$ Property: __________
4. ____ $\times 36 = 0$ Property: __________
5. $47 \times$ ____ $= 59 \times 47$ Property: __________
6. ____ $\times 1 = 102$ Property: __________

Work with a Partner

You need two dice to play this game. Take turns rolling dice. For each turn, find the product of the numbers showing for your "score." Repeat for five complete rounds. Add the five scores together. The player with the higher score wins.

TIP

When multiplying the numbers 1 through 10 by 9, the sum of the two digits is the product 9.

$6 \times 9 = \underline{54}$

$(5 + 4 = 9)$

Think About It

How are addition and multiplication alike? When would you use multiplication instead of addition?

....

....

Strategy 3: Use Place Value to Multiply

To set up a multiplication problem using whole numbers, line up the digits in a column by place value. The place values of the digits you are multiplying will tell you where to write the digits of the product.

For help understanding place value, review page 8.

Example Multiply 43 by 2.

Step 1: Align digits by place value.

	tens	ones
	4	3
×		2

Step 2: Write the product of 2 times 3 ones in the ones place.

	tens	ones
	4	3
×		2
		6

Step 3: Write the product of 2 times 4 tens in the tens place.

	tens	ones
	4	3
×		2
	8	6

$43 \times 2 = 86$

Example Multiply 102 by 4.

Step 1: Line up digits by place value.

hundreds	tens	ones
1	0	2
×		4

Step 2: Write the product of 4 times 2 ones in the ones place.

hundreds	tens	ones
1	0	2
×		4
		8

Step 3: Write the product of 4 times 0 tens in the tens place.

hundreds	tens	ones
1	0	2
×		4
	0	8

Step 4: Write the product of 4 times 1 hundred in the hundreds place.

hundreds	tens	ones
1	0	2
×		4
4	0	8

102 × 4 = 408

Try It Yourself

Write each multiplication problem in a column and solve. Check your answers on page 47.

1. 21 × 3 **2.** 11 × 9 **3.** 24 × 2 **4.** 20 × 4 **5.** 30 × 3

```
  21
×  3
  63
```

For help understanding regrouping, review page 16.

Multiplication with Regrouping

What happens when the product of two digits is greater than 9?

Example Multiply 26 by 3.

Step 1: Multiply 3 times 6 ones.

tens	ones
2	6
×	3
	18?

Step 2: Regroup 18 ones as 1 ten + 8 ones.

tens	ones
1	
2	6
×	3
	8

Step 3: Multiply 3 times 2 tens. Add the regrouped 1 ten. (3 × 2) + 1 = 7.

tens	ones
1	
2	6
×	3
7	8

26 × 3 = 78

Sometimes, you will need to regroup more than once.

Example Multiply 126 × 7.

Step 1: Multiply 7 times 6 ones.

hundreds	tens	ones
1	2	6
×		7
		42?

Step 2: Regroup 42 ones as 4 tens + 2 ones.

hundreds	tens	ones
	4	
1	2	6
×		7
		2

Step 3: Multiply 7 times 2 tens. Add the regrouped 4 tens. (7 × 2) + 4 = 18.

hundreds	tens	ones
	4	
1	2	6
×		7
	18?	2

Step 4: Regroup 18 tens as 1 hundred + 8 tens.

hundreds	tens	ones
1	4	
1	2	6
×		7
	8	2

Step 5: Multiply 7 times 1 hundred. Add the regrouped 1 hundred. (7 × 1) + 1 = 8.

hundreds	tens	ones
1	4	
1	2	6
×		7
8	8	2

126 × 7 = 882

Example Multiply 508 × 9.

Step 1: Multiply 9 times 8 ones.

hundreds	tens	ones
5	0	8
×		9
		72?

Step 2: Regroup 72 ones as 7 tens + 2 ones.

hundreds	tens	ones
	7	
5	0	8
×		9
		2

Step 3: Multiply 9 times 0 tens. Add the regrouped 7 tens. (9 × 0) + 7 = 7.

hundreds	tens	ones
	7	
5	0	8
×		9
	7	2

Step 4: Multiply 9 times 5 hundreds.

hundreds	tens	ones
	7	
5	0	8
×		9
45?	7	2

Step 5: Regroup 45 hundreds as 4 thousands and 5 hundreds.

thousands	hundreds	tens	ones
4		7	
	5	0	8
×			9
	5	7	2

Step 6: Multiply 9 times 0 thousands. Add the regrouped 4 thousands. (9 × 0) + 4 = 4.

thousands	hundreds	tens	ones
4		7	
	5	0	8
×			9
4	5	7	2

508 × 9 = 4,572

TIP

In multiplication problems with 5, the product always ends in 0 or 5.

5 × 2 = 10

5 × 5 = 25

Try It Yourself

Write each multiplication problem in a column and solve. Check your answers on page 47.

1. 14 × 6 **2.** 71 ×5 **3.** 538 × 4 **4.** 209 × 3 **5.** 806 × 8

```
  2
 14
× 6
 84
```

Start Smart Challenge

Try multiplying 4,050,678 by 3.

Strategy 4: Look for Clues in Multiplication Word Problems

Some word problems give you a *rate* (cost per pound, miles per gallon, price for each, etc.) and ask you for a *total amount* based on that rate. You can use multiplication to solve this kind of word problem.

Example Jason buys 4 pounds of cherries for **$2 per pound.** What is the **total amount** Jason spends?

Math Riddle

When you multiply it by 3, the product is 15. When you multiply it by 6, the product is 30. What's the number?

The number is 5.

The problem asks for a total amount based on a given rate. This is a clue that you may need to use multiplication to find the answer. To solve the problem, multiply the rate, or the cost of the cherries ($2 per pound) times the number of pounds of cherries Jason buys (4 pounds).

$2 \times 4 = 8$ Jason spends a total of $8.

Try It Yourself

Write and solve a multiplication number sentence to answer each problem. Check your answers on page 47.

1. Pat earns $8 per hour. If she works 11 hours during one week, what is her total salary for the week? **8 × 11 = 88; Pat earns $88 for the week.**

2. Rudy's car gets 23 miles per gallon of gasoline. The car's tank holds 9 gallons. How far can Rudy drive on one tank of gasoline?

....

....

3. Kim has 3 children. The registration fee for a youth soccer league is $18 per child. How much will Kim have to pay to register her children for soccer? ....

....

4. Janine drives for 5 hours, traveling at 65 miles per hour. How many miles does she drive? ....

....

Work With a Partner

Re-read item #4 from page 35. Suppose Janine drives for 6 more hours. With your partner, list the steps you could use to find the number of miles Janine drives in 11 hours. Share your steps with the class.

Think About It

How do rates like *price per pound* help you? How do they challenge you?

....

....

TIP

Don't count zeros that are not ending zeros. For example, 30,500 has two ending zeros, NOT three.

Strategy 5: Count Zeros to Do Mental Math

Mentally multiplying by numbers that **end in one or more zeros** is easy to do.

Example Multiply 1,200 by 30.

Step 1: Count the total number of ending zeros.	Step 2: Multiply the digits that are not ending zeros.	Step 3: Add the total number of ending zeros to the product.
1200 has 2 ending zeros. 30 has 1 ending zero. 1,200 and 30 have a total of 3 ending zeros.	$12 \times 3 = 36$	36,000

1,200 × 30 = 36,000

Try It Yourself

Find each product using mental math. Check your answers on page 47.

1. 50 × 70 3,500 **2.** 600 × 10 **3.** 11,000 × 4 **4.** 9,000 × 800 **5.** 34,100 × 20

Think About It

How have you been successful using mental math?

....

....

Strategy 6: Estimate to Check Products

After you multiply, always check that the product is reasonable. One way to check a multiplication answer is to round the numbers you multiplied and then estimate the product. If your calculated product and the estimate are close, your product is probably reasonable.

For help understanding rounding, review page 13.

Example Kathy earns $296 per week. She calculates that in 3 weeks she will earn $888. Is Kathy's product reasonable?

TIP
In multiplication problems with 10, the product always ends in 0.

2 × 10 = 20

5 × 10 = 50

Estimated Product	**Compare to Calculated Product**
$296 rounds to $300	$888 is close to $900
300 × 3 = 900	$888 is a reasonable product.

Try It Yourself

1. Kerry makes $970 per month. She calculates that she will make $5,420 in 6 months. Is Kerry's product reasonable? Explain.

....

....

2. Harry has 8 payments of $309 each left to make on his car loan. He calculates that he has $2,472 left to pay. Is Harry's product reasonable?

Explain. ....

....

Check your answers on pages 47 and 48.

Think About It

How might you teach a coworker to check a calculated product by estimating?

....

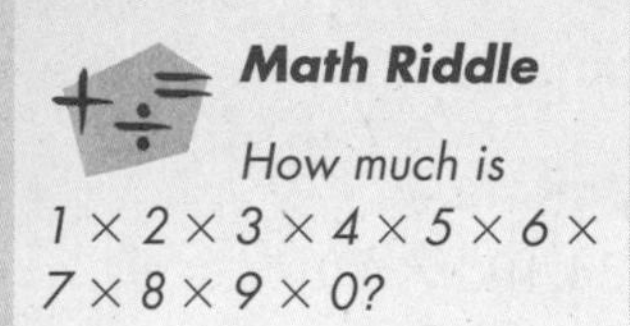

Math Riddle

How much is 1 × 2 × 3 × 4 × 5 × 6 × 7 × 8 × 9 × 0?

The answer is 0. (Zero Property)

When You Multiply Numbers

- create models of multiplication problems
- apply the properties of multiplication
- align digits by place value and regroup necessary
- find clues to choose multiplication to solve word problems
- count ending zeros to do mental math
- estimate to check products

Journal

In your journal, finish the following sentences.

- I want to become better at math so I can . . .
- My best math experience was when . . .
- When I hear someone say math is fun, I . . .

Next, look at the goals you set on page 7. Note your progress for each of your goals. If you have achieved a goal, put a check next to it and congratulate yourself!!

Topic 5: **Division**

In this section, you will learn how to:

- ***model division problems***
- ***recognize and apply properties of division***
- ***divide numbers and regroup amounts as necessary***
- ***account for remainders when dividing***
- ***solve word problems with division***
- ***check quotients using multiplication***

When you split the bill after lunch with a friend, you are using division. Have you ever calculated how many payments you have left on a loan? That's division, too. There are many ways we use division on a regular basis.

division *splitting a single amount into a number of equal amounts*

quotient *the answer to a division problem*

Strategy 1: **Understand Division Problems**

One way to model division is to think of **creating equal-sized groups** out of a whole. If you look at it this way, the expression "10 ÷ 5" means "how many groups of 5 can you make from 10?" You can use a picture to show what is going on in a division problem and to find the quotient.

Example Use a picture to model 10 ÷ 5 and find the quotient.

10 ÷ 5 =

How many groups of 5 can you make? 10 ÷ 5 = 2

You also can think of division as **repeated subtraction.** Then, the expression "10 ÷ 5" means "how many times do you have to subtract 5 from 10 to get to 0?" You can use a number line to show a division problem and to find the quotient.

Example Use a number line to model 10 ÷ 5 and find the quotient.

10 ÷ 5 =

How many times did you have to subtract 5? 10 ÷ 5 = 2

For help understanding number lines, review page 11.

Try It Yourself

1. Create a picture to model 8 ÷ 2 and find the quotient.

2. Create a number line to model 9 ÷ 3 and find the product.

Check your answers on page 48.

Work with a Group

Work together to solve this math riddle. *I am 5 digits long. I am divisible by 3 and 9, but not 6. My digits add up to 27. When my first and last digits are added, you will get a multiple of seven. My lowest digit is 3. If you add up my second and fourth digits, you get 10. No number is used more than once. On the left of the comma, the numbers from left to right decrease by one. On the right of the comma, the numbers increase by three.* Check your answer on page 48.

Think About It

Which kind of model helps you most to understand division? Why?

....

....

Strategy 2: Recognize and Apply Division Properties

Here are four facts about division that you need to know.

1. The Identity Property

Dividing a number by 1 does not change the value of the number.

Example $3 \div 1 = 3$

How many groups of 1 can you make from 3?

3 groups

How many times do you have to subtract 1 from 3 to get to 0?
3 times

$3 \div 1 = 3$

2. The Division by the Same Number Property

Dividing a number by itself (except 0) gives a quotient equal to 1.

Example $3 \div 3 = 1$

How many groups of 3 can you make from 3?

1 group

How many times do you have to subtract 3 from 3 to get to 0?
1 time

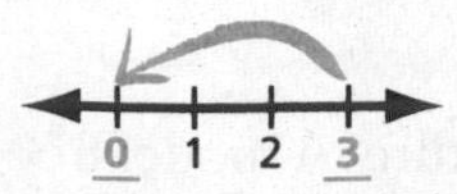

$3 \div 3 = 1$

3. The Zero Property
Dividing 0 by any number (except 0) gives a quotient equal to 0.

Example $0 \div 3 = 0$

How many groups of 3 can you make from 0?
None

How many times do you have to subtract 3 from 0 to get to 0
None

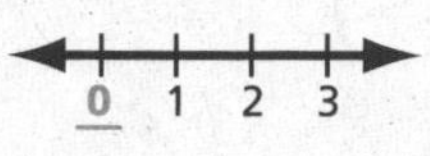

$0 \div 3 = 0$

Math Riddle

When things go wrong, what can you always count on?

Your fingers.

4. The Division by Zero Property
Division by zero is undefined. You cannot divide by 0.

Example $3 \div 0$ has no answer.

What happens when you try to split 3 up into 0 groups? You can't do it! If there aren't any groups, then you can't put the 3 anywhere. To make this idea clear, mathematicians call division by zero "undefined"—meaning that it doesn't really make any sense to try to split a number into zero groups.

Try It Yourself

Apply the correct property of division to make each number sentence true. State which property you use. Check your answers on page 48.

1. $0 \div 5 =$ 0 Property: zero

2. $7 \div$ ____ = undefined Property: ____________

3. ____ $\div 46 = 1$ Property: ____________

4. $19 \div$ ____ $= 19$ Property: ____________

5. $238 \div$ ____ $= 1$ Property: ____________

6. $85 \div 0 =$ ____ Property: ____________

7. ____ $\div 1 = 99$ Property: ____________

8. ____ $\div 2 = 0$ Property: ____________

Think About It
Why is division sometimes called repeated subtraction?

....

....

Strategy 3: Use Place Value to Divide

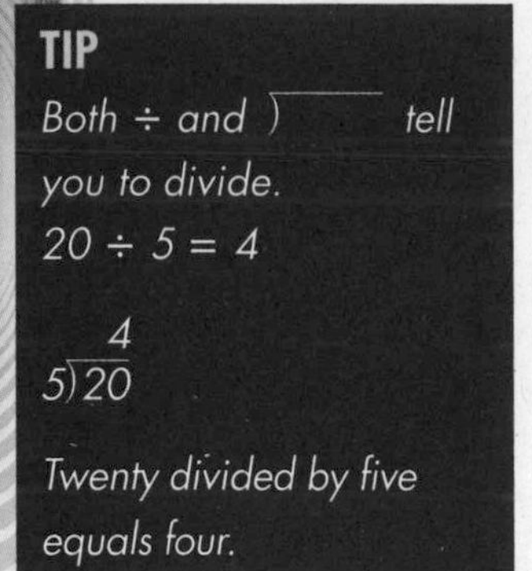

To set up a division problem, use the division symbol.

The place values of the digits you are dividing will tell you where to write the digits of the quotient.

Write the number you are dividing by to the left.

Write the number being divided under the bar

Example 28 ÷ 2

Step 1: Write the problem using the division symbol.

$2\overline{)28}$

Step 2: Divide <u>2 tens by 2</u>. Write the quotient in the tens place.

tens ones

$$\begin{array}{r} 1 \\ 2\overline{)28} \end{array}$$

Step 3: Multiply 1 ten by 2 and subtract the product from 2 tens.

tens ones

$$\begin{array}{r} 1 \\ 2\overline{)28} \\ \underline{-2} \\ 0 \end{array}$$

Step 4: Bring down the ones.

tens ones

$$\begin{array}{r} 1 \\ 2\overline{)28} \\ \underline{-2} \\ 08 \end{array}$$

Step 5: Divide <u>8 ones by 2</u>. Write the quotient in the ones place.

tens ones

$$\begin{array}{r} 14 \\ 2\overline{)28} \\ \underline{-2} \\ 08 \end{array}$$

Step 6: Multiply 4 ones by 2 and subtract the product from 8 ones.

tens ones

$$\begin{array}{r} 14 \\ 2\overline{)28} \\ \underline{-2} \\ 08 \\ \underline{-8} \\ 0 \end{array}$$

28 ÷ 2 = 14

Try It Yourself

Write each division problem using the division symbol and solve. Check your answers on page 48.

TIP
Circle the number of the problem, so you don't think it's part of the problem.

1. 84 ÷ 4 **2.** 36 ÷ 3 **3.** 55 ÷ 5 **4.** 62 ÷ 2 **5.** 93 ÷ 3

$$\begin{array}{r} 21 \\ 4\overline{)84} \\ \underline{-8} \\ 04 \\ \underline{-4} \\ 0 \end{array}$$

Division with Regrouping

What happens when a digit you want to divide is less than the number you are dividing by?

Example Divide 246 by 3.

Step 1: You cannot divide the hundreds because 2 < 3.

100s 10s 1s

3)2 4 6

Step 2: Regroup all the hundreds as tens. 20 tens + 4 tens = 24 tens

100s 10s 1s

3)0 24 6

Step 3: Divide the tens.

100s 10s 1s

```
   8
3)0 24 6
  −24
    0
```

Step 4: Bring down the ones.

100s 10s 1s

```
   8
3)0 24 6
  −24
    0 6
```

Step 5: Divide the ones.

100s 10s 1s

```
   8 2
3)0 24 6
  −24
    0 6
    − 6
      0
```

For help understanding regrouping, review page 16.

246 ÷ 3 = 82

Try It Yourself

Write each division problem using the division symbol and solve. Check your answers on page 48.

1. 147 ÷ 7 **2.** 249 ÷ 3 **3.** 305 ÷ 5 **4.** 328 ÷ 4 **5.** 366 ÷ 6

```
   2 1
7)0 14 7
  −14
    0 7
    − 7
      0
```

Math Riddle

It is the product of 6 and 8. When you divide it by 4, the quotient is 12. What's the number?

The number is 48.

Start Smart Challenge

Try dividing 5,445 by 9.

Strategy 4: Account for Remainders

Sometimes, you cannot split an amount into equal groups. The amount "left over" after you divide is called the **remainder.**

Example Divide 5 by 3.

How many groups of 3 can you make from 5? What amount is left over?
1 group, with 2 left over.

How many times can you subtract 3 from 5 **without going past 0?**
1 group, with 2 left over.

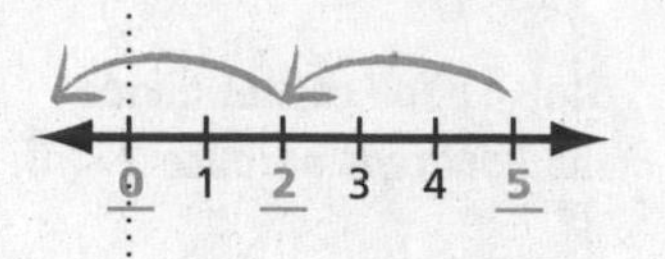

5 ÷ 3 = 1 R2. The letter *R* means "remainder."

Example Use the division symbol to divide 56 by 3.

Step 1: Write the problem using the division symbol.

$3\overline{)56}$

Step 2: Divide 5 tens by 3. (Think "3 times what number is closest to, but still less than, 5?')

tens ones

$$\begin{array}{r} 1 \\ 3\overline{)56} \\ -3 \\ \hline 2 \end{array}$$

Step 3: Bring down the ones.

tens ones

$$\begin{array}{r} 1 \\ 3\overline{)56} \\ -3 \\ \hline 26 \end{array}$$

Step 4: Divide 26 ones by 3. (Think "3 times what number is closest to, but still less than, 26?')

tens ones

$$\begin{array}{r} 18 \\ 3\overline{)56} \\ -3 \\ \hline 26 \\ -24 \\ \hline 2 \end{array}$$

Step 5: Anything left over after you have divided the ones is a remainder.

tens ones

$$\begin{array}{r} 18 \\ 3\overline{)56} \\ -3 \\ \hline 26 \\ -24 \\ \hline 2 \end{array}$$

56 ÷ 3 = 18 R2

Try It Yourself

Rewrite each division problem and solve. Check your answers on page 48.

1. 91 ÷ 8 **2.** 75 ÷ 2 **3.** 359 ÷ 7 **4.** 226 ÷ 5 **5.** 604 ÷ 7

$$\begin{array}{r} 11\text{, R3} \\ 8\overline{)91} \\ -8 \\ \hline 11 \\ -8 \\ \hline 3 \end{array}$$

Think About It

If you have to split a pizza evenly among 4 children, which would you rather have: a large pizza cut into 10 slices or a large pizza cut into 12 slices?

....

....

Strategy 5: Look for Clues in Division Word Problems

Some word problems ask you to find a *rate* (cost per pound, price for each, amount per person, etc.). You can use division to solve this kind of problem.

TIP

Read word problems several times, if that helps. Draw a picture or take notes to help you "see" what to do.

Example John works 8 hours and earns $72. How much is John earning per hour?

The problem asks for a rate (dollars per hour). This is a clue telling you that you may need to use division to find the answer. To solve the problem, divide the total amount of money John earns ($72) by the number of hours John works (8 hours).

$72 \div 8 = 9$ John earns $9 per hour.

Try It Yourself

Write and solve a division number sentence to answer each problem. Check your answers on page 48.

1. Dana is setting up 63 chairs in a meeting room. If she puts the chairs in 9 equal rows, how many chairs will she use per row?

63 ÷ 9 = 7; Dana will use 7 chairs per row.

2. Audra owes $1,743 on her car loan. If she has 7 equal payments left, what is the amount of each payment?

....

....

3. What is the hourly labor charge for a mechanic if she charges $288 for 6 hours labor?

....

....

Work With a Partner

Create a short word problem that a student could solve with division. The solution should be $6. Share your word problem with the class.

Think About It

Suppose you divided 15 by 2 and got 6 remainder 3. What does it mean when a remainder is greater than the number you are dividing by?

Strategy 6: Use Multiplication to Check a Quotient

You can check a quotient with multiplication because multiplication and division are inverse operations.

Example Bethany calculates $52 \div 4 = 13$. Is this quotient correct?

She started at 52 and got to 13 by dividing by 4.	**Can Bethany start at 13 and get back to 52 by multiplying by 4?**
$\begin{array}{r} 13 \\ 4\overline{)52} \\ -4 \\ \hline 12 \\ -12 \\ \hline 0 \end{array}$	$\begin{array}{r} {}^{1} \\ 13 \\ \times\ 4 \\ \hline 52 \end{array}$ ⟵ Yes. the quotient is correct.

Try It Yourself

Use multiplication to check each quotient.

1. $256 \div 8 = 32$? **2.** $198 \div 9 = 22$? **3.** $85 \div 5 = 15$? **4.** $246 \div 6 = 41$?

yes $\begin{array}{r} {}^{1} \\ 32 \\ \times\ 8 \\ \hline 256 \end{array}$

Think About It

How can you use multiplication to help you divide?

When You Divide Numbers

- Create models of division problems.
- Apply the properties of division.
- Line digits by place value and regroup when necessary.
- Account for remainders.
- Identify clues that tell you to divide in word problems.
- Multiply to check quotients.

Math Riddle

Find three even numbers that follow each other and divide evenly into 12.

2; 4; 6

Journal

In your journal, describe how you used math this week (outside of class).

Next, look at the goals you set on page 7. Note your progress for each of your goals. If you have achieved a goal, put a check next to it and congratulate yourself!

What Works for You?

Are your math habits changing? Review the math habits you marked on page 5 and answer the questions.

How have my math habits changed?
What math habits do I want to work on?

Look at goals you set on page 7. Have you met your goals? Answer the questions.

Which goals have I achieved so far?
Which goals do I want to work on?

Journal

In your journal, explain how you feel about math now as compared to before you finished this book.

Answers

Page 9 Try It Yourself

2. tens, 50
3. hundreds, 500
4. thousands, 5,000
5. 21
6. thirteen thousand
7. sixty-five
8. one thousand, three hundred forty
9. 25
10. 99
11. 307
12. 1,120

Page 10 Try It Yourself

2. multiply, times
3. ÷
4. equals, is
5. is less than
6. >
7. Nine plus one equals ten.
8. $3 \times 5 = 15$
9. Six minus two equals four.
10. Forty-five does not equal forty-six.

Page 11 Try It Yourself

2. left of zero
3. right of zero

Page 12 Try It Yourself

2. <
3. >
4. <
5. <
6. >

Page 13 Try It Yourself

2. tens; 670
3. hundreds; 300
4. thousands; 13,000

Page 15 Try It Yourself

2. 91; identity
3. 7; associative
4. 0; identity
5. 40; commutative
6. 67; associative

Page 16 Try It Yourself

2. 27
3. 899
4. 886
5. 58

Page 16 Work with a Group

936

Page 18 Try It Yourself

2.
$$\begin{array}{rrrr} & 1 & & \\ & 3 & 9 & 2 \\ + & & 8 & 4 \\ \hline & 4 & 7 & 6 \end{array}$$

3.
$$\begin{array}{rrr} & 1 & 1 \\ & 6 & 4 \\ + & 5 & 9 \\ \hline 1 & 2 & 3 \end{array}$$

4.
$$\begin{array}{rrrr} & 1 & 1 & \\ & 4 & 8 & 0 \\ + & 5 & 6 & 6 \\ \hline 1, & 0 & 4 & 6 \end{array}$$

Page 18 Start Smart Challenge

$$\begin{array}{rrrrr} 1 & 1 & 1 & 1 & \\ & 9, & 1 & 9 & 1 \\ + & 1, & 9 & 1 & 9 \\ \hline 1 & 1, & 1 & 1 & 0 \end{array}$$

Page 18 Try It Yourself

2. circle *in all;*

$$\begin{array}{rrr} & 1 & \\ & 1 & 3 \\ + & & 8 \\ \hline & 2 & 1; \end{array}$$

Ramona typed 21 pages.

3. circle *combined;*

$$\begin{array}{rrr} & 1 & \\ & 2 & 5 \\ + & 2 & 5 \\ \hline & 5 & 0; \end{array}$$

Curtis lifts 50 pounds.

Page 19 Try It Yourself

2. 60
3. 90
4. 80
5. 220

Page 20 Try It Yourself

1. Yes; 39 rounds to 40 and 23 rounds to 20. $40 + 20 = 60$, which is close to 62.
2. No; 12 rounds to 10 and 19 rounds to 20. $10 + 20 = 30$, which is not close to 51.

Page 21 Try It Yourself

2. 2; zero
3. 8; zero
4. 0; identity

Page 23 Try It Yourself

2.
$$\begin{array}{rrr} & 2 & 1 \\ - & 1 & 0 \\ \hline & 1 & 1 \end{array}$$

3.
$$\begin{array}{rrr} & 7 & 9 \\ - & 5 & 5 \\ \hline & 2 & 4 \end{array}$$

4.
$$\begin{array}{rrrr} & 8 & 4 & 1 \\ - & & 4 & 1 \\ \hline & 8 & 0 & 0 \end{array}$$

5.
$$\begin{array}{rrrr} & 9 & 7 & 6 \\ - & 3 & 2 & 4 \\ \hline & 6 & 5 & 2 \end{array}$$

Page 25 Try It Yourself

2.
$$\begin{array}{rrrr} & 3 & 13 & \\ & \cancel{4} & \cancel{3} & 6 \\ - & & 8 & 1 \\ \hline & 3 & 5 & 5 \end{array}$$

3.
$$\begin{array}{rrrr} & 0 & 11 & \\ & \cancel{1} & \cancel{1} & 5 \\ - & & 9 & 2 \\ \hline & & 2 & 3 \end{array}$$

4.
$$\begin{array}{rrrr} & & 12 & \\ & 2 & \cancel{2} & 12 \\ & \cancel{3} & \cancel{3} & \cancel{2} \\ - & 1 & 9 & 9 \\ \hline & 1 & 3 & 3 \end{array}$$

5.
$$\begin{array}{rrrr} & & 14 & \\ & 5 & \cancel{4} & 11 \\ & \cancel{6} & \cancel{5} & \cancel{1} \\ - & 2 & 7 & 7 \\ \hline & 3 & 7 & 4 \end{array}$$

6.
$$\begin{array}{r} 9\ \ \ \\ 3\ \cancel{10}\ 10 \\ \cancel{4}\ \cancel{0}\ \cancel{0} \\ -\quad 3\ 5 \\ \hline 3\ 6\ 5 \end{array}$$

7.
$$\begin{array}{r} 9\ \ \ \\ 7\ \cancel{10}\ 15 \\ \cancel{8}\ \cancel{0}\ \cancel{5} \\ -\quad\ \ 7 \\ \hline 7\ 9\ 8 \end{array}$$

8.
$$\begin{array}{r} 4\ 10\ \ \ \\ \cancel{5}\ \cancel{0}\ 3 \\ -\quad 7\ 3 \\ \hline 4\ 3\ 0 \end{array}$$

Page 25 Start Smart Challenge

1,234,567

Pages 25–26 Try It Yourself

2. circle *left over;*

$$\begin{array}{r} 1\ 14 \\ \cancel{2}\ \cancel{4} \\ -\ 1\ 6 \\ \hline 8 \end{array};$$

Melanie has 8 ounces of tomatoes left over.

3. circle *how much less;*

$$\begin{array}{r} 1\ 7\ 5 \\ -\ 1\ 5\ 0 \\ \hline 2\ 5 \end{array};$$

Alfredo weighs 25 pounds less.

4. circle *remain;*

$$\begin{array}{r} 9\ \ \ \\ 0\ \cancel{10}\ 15 \\ \cancel{1}\ \cancel{0}\ \cancel{5} \\ -\quad 5\ 9 \\ \hline 4\ 6 \end{array};$$

46 minutes remain.

Page 27 Try It Yourself

2. 36
3. 19
4. 19
5. 36

Page 28 Try It Yourself

1. Yes; 215 rounds to 220 and 187 rounds to 190. $220 - 190 = 30$, which is close to 28.
2. No; 476 rounds to 480 and 603 rounds to 600. $600 - 480 = 120$, which is not close to 157.
4. 9 is correct; $9 + 85 = 94$
5. 33 is incorrect; $33 + 38 \neq 75$; the correct answer is 37.
6. 139 is correct; $139 + 261 = 400$
7. 288 is incorrect; $288 + 72 \neq 350$; the correct answer is 278.

Page 29 Try It Yourself

1. (■■) + (■■) + (■■) + (■■); $2 \times 4 = 8$
2. 0 1 2 3 4 5 6 7 8 9 10 11 12 13 14 15 16 17 18 19 20; $3 \times 6 = 18$

Page 31 Try It Yourself

2. 0; zero
3. 1; identity
4. 0; zero
5. 59; commutative
6. 102; identity

Page 32 Try It Yourself

2. $\begin{array}{r} 1\ 1 \\ \times\ \ 9 \\ \hline 9\ 9 \end{array}$

3. $\begin{array}{r} 2\ 4 \\ \times\ \ 2 \\ \hline 4\ 8 \end{array}$

4. $\begin{array}{r} 2\ 0 \\ \times\ \ 4 \\ \hline 8\ 0 \end{array}$

5. $\begin{array}{r} 3\ 0 \\ \times\ \ 3 \\ \hline 9\ 0 \end{array}$

Page 33 Try It Yourself

2. $\begin{array}{r} 3\ \ \ \ \ \\ 7\ 1 \\ \times\quad 5 \\ \hline 3\ 5\ 5 \end{array}$

3. $\begin{array}{r} 2\ 1\ 3\ \ \\ 5\ 3\ 8 \\ \times\quad\ \ 4 \\ \hline 2,1\ 5\ 2 \end{array}$

4. $\begin{array}{r} 2\ \ \\ 2\ 0\ 9 \\ \times\quad\ \ 3 \\ \hline 6\ 2\ 7 \end{array}$

5. $\begin{array}{r} 6\ \ \ \ 4\ \ \\ 8\ 0\ 6 \\ \times\quad\ \ 8 \\ \hline 6,4\ 4\ 8 \end{array}$

Page 33 Start Smart Challenge

$$\begin{array}{r} 1\ \ \ \ 2\ \ 2\ 2\ \ \ \ \ \\ 4\ 0\ 5\ 0\ \ 6\ 7\ 8 \\ \times\qquad\qquad\quad 3 \\ \hline 1\ 2,1\ 5\ 2,0\ 3\ 4 \end{array}$$

Page 34 Try It Yourself

2. $23 \times 9 = 207$; Rudy can drive 207 miles.
3. $18 \times 3 = 54$; Kim will have to pay $54.
4. $65 \times 5 = 325$; Janine drives 325 miles.

Page 35 Work With a Partner

Janine drives 715 miles in 11 hours. Lists of steps will vary.

Page 35 Try It Yourself

2. 6,000
3. 44,000
4. 7,200,000
5. 682,000

Page 36 Try It Yourself

1. No; 970 rounds to 1,000. $6 \times 1{,}000 = 6{,}000$, which is not close to 5,420.

2. Yes; 309 rounds to 300. $8 \times 300 = 2{,}400$, which is close to 2,472.

Page 37 Try It Yourself

1. ; $8 \div 2 = 4$
2. 0 1 2 3 4 5 6 7 8 9 ; $9 \div 3 = 3$

Page 38 Work with a Group

54, 369

Page 39 Try It Yourself

2. 0; division by zero
3. 46; division by same number
4. 1; identity
5. 238; division by same number
6. undefined; division by zero
7. 99; identity
8. 0; zero

Page 40 Try It Yourself

2. $3\overline{)36}$ = 12
   −3
   06
   −6
   0
3. $5\overline{)55}$ = 11
   −5
   05
   − 5
   0
4. $2\overline{)62}$ = 31
   −6
   02
   − 2
   0
5. $3\overline{)93}$ = 31
   −9
   03
   − 3
   0

Page 41 Try It Yourself

2. $3\overline{)0\,24\,9}$ = 8 3
   −24
   0 9
   − 9
   0
3. $5\overline{)0\,30\,5}$ = 6 1
   −30
   0 5
   − 5
   0
4. $4\overline{)0\,32\,8}$ = 8 2
   −32
   0 8
   − 8
   0
5. $6\overline{)0\,36\,6}$ = 6 1
   −36
   0 6
   − 6
   0

Page 41 Start Smart Challenge

$9\overline{)0\,54\,45}$ = 6 05
−54
0 45
−45
0

Page 42 Try It Yourself

2. 37 R1
   $2\overline{)7\,5}$ = 3 7
   −6
   1 5
   −1 4
   1
3. 51 R2
   $7\overline{)0\,35\,9}$ = 5 1
   −35
   0 9
   − 7
   2
4. 45 R1
   $5\overline{)0\,22\,6}$ = 4 5
   −20
   2 6
   −2 5
   1
5. 86 R2
   $7\overline{)0\,60\,4}$ = 8 6
   −56
   4 4
   −4 2
   2

Page 43 Try It Yourself

2. $1{,}743 \div 7 = 249$; Each payment is $249.
3. $288 \div 6 = 48$; The hourly labor charge is $48 per hour.

Page 44 Try It Yourself

2. yes
   1
   2 2
   × 9
   1 9 8
3. no
   2
   1 5
   × 5
   7 5
4. yes
   4 1
   × 6
   2 4 6